Beginners Guide to UPLC

Ultra-Performance Liquid Chromatography

Eric S. Grumbach, Joseph C. Arsenault, Douglas R. McCabe

Waters
THE SCIENCE OF WHAT'S POSSIBLE.™

Waters Corporation
34 Maple Street
Milford, MA 01757

Library of Congress Control Number: 2009930076

Printed in the USA
June 2009

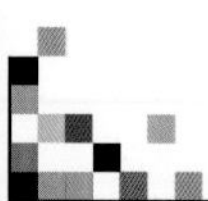

Acknowledgments

The authors would like to acknowledge the following colleagues from Waters Corporation for their contributions to this manuscript: Dr. Mark Baynham, Kenneth Fountain, Ian Hanslope, Dawn Maheu, Dr. Patrick McDonald, Damian Morrison, Dr. Uwe Neue and Paul Rainville.

In Memorium

We would like to dedicate this publication to our colleague, and great friend, William Martin David Collis (Dave). Dave had been a Waters employee for over 25 years. He was involved in the development of educational and training materials for Waters customers and personnel, and was instrumental during the successful world-wide launch of Waters ACQUITY UPLC Technology, in 2004. His command of the science of chromatography, as well as presentation style, made him a featured speaker around the world. Dave passed away during the production of this Primer, and our thoughts and prayers go out to his family. He will be missed.

Dave Collis

15th April 1957 - 20th May 2009

Table of Contents

List of Figures

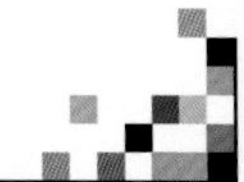

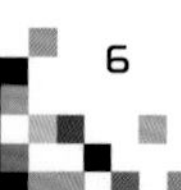

Introduction

In 2004, separation science was revolutionized with the introduction of Ultra-Performance Liquid Chromatography [UPLC® Technology]. Significant advances in instrumentation and column technology were made to achieve dramatic increases in resolution, speed and sensitivity in liquid chromatography. For the first time, a holistic approach involving simultaneous innovations in particle technology and instrument design was endeavored to meet and overcome the challenges of the analytical laboratory. This was done in order to make analytical scientists more successful and businesses more profitable and productive.

For more than four decades, reducing stationary-phase particle size has been exploited to improve chromatographic separation efficiency. Until recently, LC technology had reached a plateau in which the benefits of reducing particle size could not be fully realized due to the negative influence of instrument band spreading and limited pressure range.

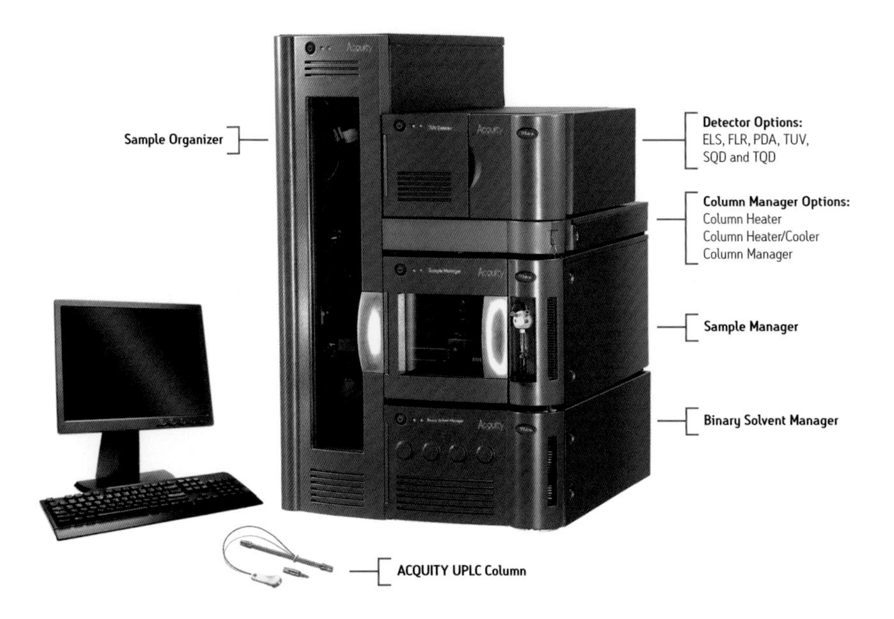

Figure 1: The ACQUITY UPLC® System.

The ACQUITY UPLC System [Figure 1] has removed those barriers, enabling columns packed with smaller particles [1.7 – 1.8 µm] to reach their theoretical performance, while precisely delivering mobile phase at pressures up to 1030 bar [15,000 psi], thus providing a new level of chromatographic performance.

UPLC Technology facilitates improvements of resolution, sensitivity and speed to be achieved, without compromise. Whether the separation goal is to achieve ultra-fast analysis, increase throughput while maintaining resolution, improving resolution while decreasing analysis time or achieve ultra-high resolution, the flexibility of the ACQUITY UPLC System enables method requirements to be met [Figure 2].

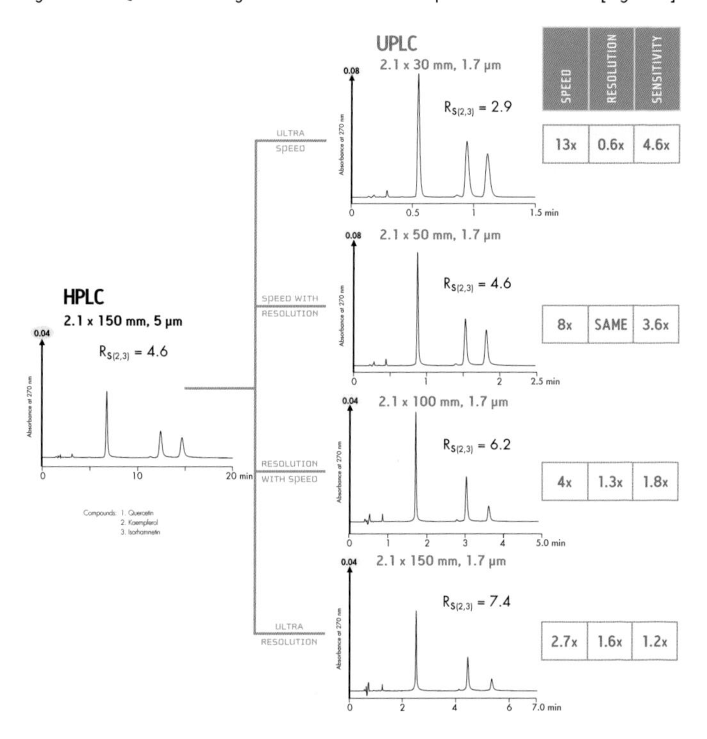

Figure 2: Versatility of UPLC Technology – achieving resolution, sensitivity and speed.

This technology primer is designed to provide new, existing and potential UPLC users the ability to understand how UPLC Technology works, how to be successful with it, and how it can provide impactful results within their organization.

Bands, Peaks and Band Spreading

How a Chromatographic Band is Formed

A sample mixture is transferred from a sample vial into a moving fluidic stream [mobile phase]. The sample is then carried by the mobile phase to the head of a chromatographic column by a high pressure pump. The mobile phase and injected sample mixture enters the column, passes through the particle bed, and exits, transferring the separated mixture to a detector [Figure 3].

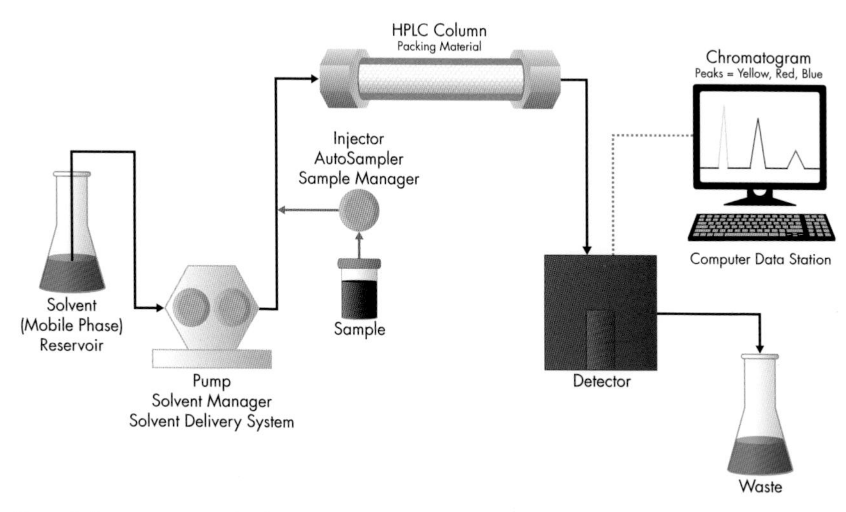

Figure 3: Representation of an HPLC system.

First, let's consider how the sample band is separated into individual analyte bands [flow direction is represented by green arrows]. Figure 4A represents the column at time zero [the moment of injection], when the sample enters the column and begins to form a band. The sample shown here, a mixture of yellow, red, and blue dyes, appears at the inlet of the column as a single black band.

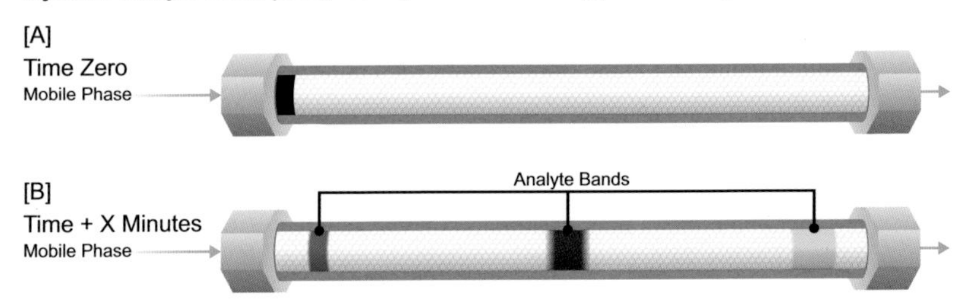

Figure 4: Understanding how a chromatographic column works – analyte bands.

After a few minutes, during which mobile phase flows continuously and steadily through the packing material particles, we can see that the individual dyes have moved in separate bands at different speeds [Figure 4B]. This is because there is a competition between the mobile phase and the stationary phase for attracting each of the dyes or analytes. Notice that the yellow dye band moves the fastest and is about to exit the column. The yellow dye has a greater affinity for [is attracted to] the mobile phase more than the other dyes. Therefore, it moves at a faster speed, closer to that of the mobile phase. The blue dye band has a greater affinity for the packing material more than the mobile phase. Its stronger attraction to the particles causes it to move significantly slower. In other words, it is the most retained compound in this sample mixture. The red dye band has an intermediate attraction for the mobile phase and therefore moves at an intermediate speed through the column. Since each dye band moves at a different speed, we are able to separate the mixture chromatographically.

How a Chromatographic Band Becomes a Peak

Each specific analyte band is made up of many analyte molecules. The center of the band contains the highest concentration of analyte molecules; while the leading and trailing edges of the band are decreasingly less concentrated as they interface with the mobile phase [Figure 5].

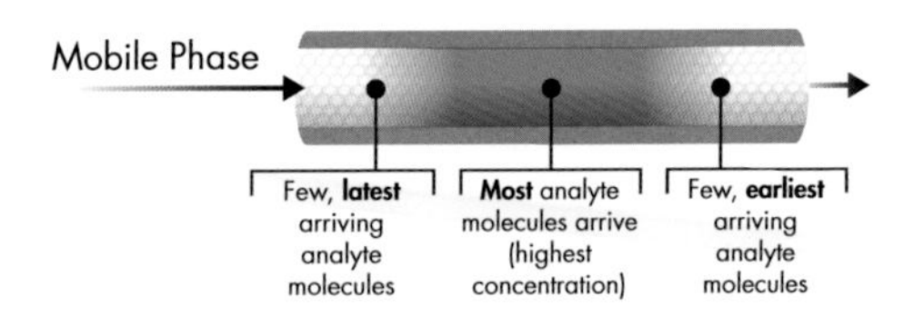

Figure 5: Concentration profile of green analyte molecules within the analyte band.

As the separated dye bands leave the column, they pass immediately into a detector. The detector sees [detects] each separated compound band against a background of mobile phase [see Figure 6]. An appropriate detector [UV, ELS, fluorescence, mass...etc.] has the ability to sense the presence of a compound and send its corresponding electrical signal to a computer data station where it is recorded as a peak. The detector responds to the varying concentration of the specific analyte molecules within the band, where the center of the band [highest population of analyte molecules] is interpreted by the detector as the apex of the peak.

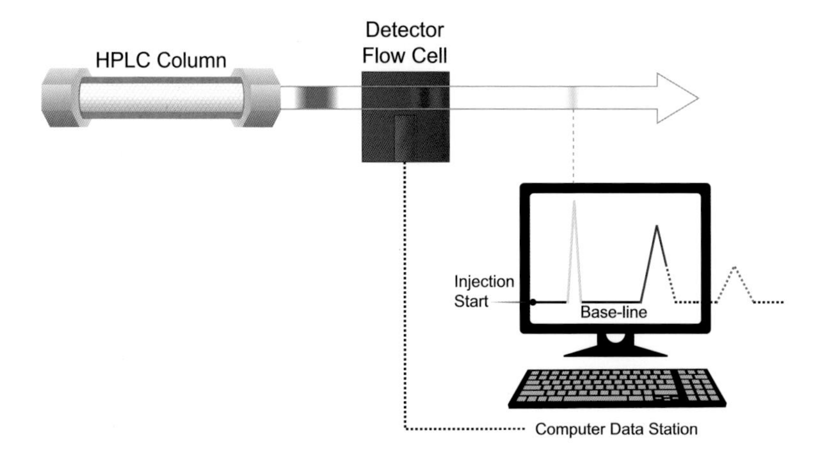

Figure 6: Peaks are digitally created as an electronic response to the analyte band as it passes through the detector.

What Is a Chromatogram?

A chromatogram is a representation of the separation that has chemically [chromatographically] occurred in the HPLC system. A series of peaks rising from a baseline is drawn on a time axis. Each peak represents the detector response for a different compound. The chromatogram is plotted by the computer data station [see Figure 6]. The yellow band has completely passed through the detector flow cell; the electrical signal generated has been sent to the computer data station. The resulting chromatogram has begun to appear on the screen. Note that the chromatogram begins when the sample is first injected and starts as a straight line set near the bottom of the screen. This is called the baseline; it represents pure mobile phase passing through the flow cell over time. As the yellow analyte band passes through the flow cell, a signal [which varies depending on the concentration of the analyte molecules] is sent to the computer. The line curves, first upward, and then downward, in proportion to the concentration of the yellow dye in the sample band. This creates a peak in the chromatogram. After the yellow band passes completely out of the detector cell, the signal level returns to the baseline; the flow cell now has, once again, only pure mobile phase in it. Since the yellow band moves fastest, eluting first from the column, it is the first peak drawn. A little while later, the red band reaches the flow cell. The signal rises up from the baseline as the red band first enters the cell, and the peak representing the red band begins to be drawn. In this diagram, the red band has not fully passed through the flow cell. The diagram shows what the red band/peak would look like if we stopped the process at this moment. Since most of the red band has passed through the cell, most of the peak has been drawn, as shown by the solid line. If we would continue the chromatographic process, the red band would completely pass through the flow cell and the red peak would be completed [dotted line]. The blue band, the most strongly retained, travels at the slowest rate and elutes after the red band. The dotted line shows you how the completed chromatogram would appear if we had let the run continue to its conclusion. It is interesting to note that the width of the blue peak will be the broadest because the width of the blue analyte band, while narrowest on the column, becomes the widest once it leaves the column. This is because it moves more slowly through the chromatographic packing material bed and requires more time [and mobile phase volume] to be eluted completely. Since mobile phase is continuously flowing at a fixed rate, this means that the blue band widens and is more dilute. The detector responds in proportion to the concentration of the band, therefore, the blue peak is lower in height and larger in width.

Band Spreading

Before the sample/analyte band reaches the detector, it will pass through multiple components of the chromatographic system that will contribute to the distortion and broadening of the chromatographic band [Figure 7]. This phenomenon is referred to as band spreading. As the analyte band becomes wider, the resulting chromatographic peak width is increased. The wider band results in a dilution effect that produces a decrease in peak height accompanied by a loss in sensitivity and resolution. Conversely, if band spreading is minimized, narrower chromatographic bands are achieved, resulting in higher efficiency. These taller, narrower chromatographic peaks are easier for the detector to see, allowing for higher sensitivity and resolution due to a more concentrated analyte band. It is therefore important to be aware of the factors that influence band spreading in efforts to reduce and control those influences to improve overall chromatographic performance.

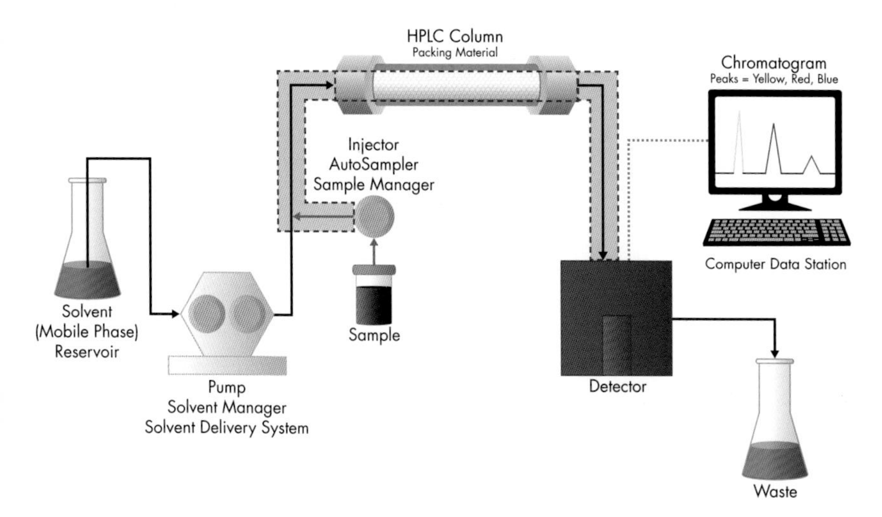

Figure 7: Band spreading will occur along the flow path from the injector (sample band), into, through and out of the column (analyte bands), and then into the detector.

Within a chromatographic system, both column and extra-column [everything outside the column] influences contribute to band spreading. Extra-column sources include the injection volume, the fluidic path of the instrument between the sample injector and the column, and from the exit of the column to the detector [including the flow cell] and all connections included therein. Column sources of band spreading include the particle size of the packing material, how well the material within the column is packed and its diffusion characteristics in relation to the speed of the mobile phase, analyte size and geometry. The summation of the variances (σ^2) for each one of these contributors impacts the width of a peak [Figure 8].

$$\underbrace{\sigma^2 \text{ variances}}_{\text{[total system band spread]}} = \underbrace{\sigma^2 \text{ extra column}}_{\text{[instrument]}} + \underbrace{\sigma^2 \text{ column}}_{\text{[intra-column]}}$$

Figure 8: As a statistical function, plate count can be thought of as the population variance [σ^2], which is a summation of both extra-column and column variances. The population variance [σ^2] measures the narrowness of a Gaussian peak [σ] relative to its elution volume.

In order to achieve significantly improved performance in liquid chromatography, a reduction in both extra-column and column band spreading influences is necessary. The ACQUITY UPLC System is based on the concept of reducing both forms of band spreading such that gainful improvements in efficiency and sensitivity can be reached [Figure 9].

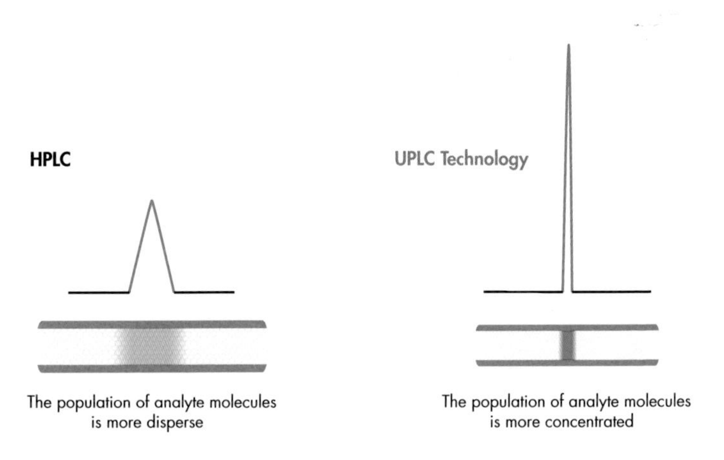

Figure 9: Impact of system band spreading [column and extra-column] on peak shape.

Addressing Extra-column [Instrument] Band Spreading

To realize a significant improvement in chromatographic performance, a substantial engineering effort was undertaken to design an LC instrument that could accommodate the pressures produced by small particle [sub-2 µm] packings, while minimizing the dispersion of the flow path such that theoretical performance of the separation column could be achieved. It is equally important that the instrument is a reliable, robust and precise analytical tool for routine laboratory use. To demonstrate the importance of instrument design, one needs to understand how system dispersion [instrument + column band spreading] can impact chromatographic results.

The measurement of plate count [N], or efficiency, takes into account the dispersion of a peak. The width of a peak is directly related to how wide the analyte band is as it passes through the detector, while the apex of that peak is the point at which the greatest concentration of analyte molecules exists within that band. This measurement is performed under isocratic conditions.

The plate count equation can be derived as shown in Figure 10, where [V_n] is the elution volume of the peak, [w] is peak width, and [a] is a constant that is determined based on the peak height at which the width of the peak is measured. Each method of measuring peak width will yield identical plate count results, as long as the peak is perfectly symmetrical. If the peak exhibits any fronting or tailing, these methods of measurement will yield different results.

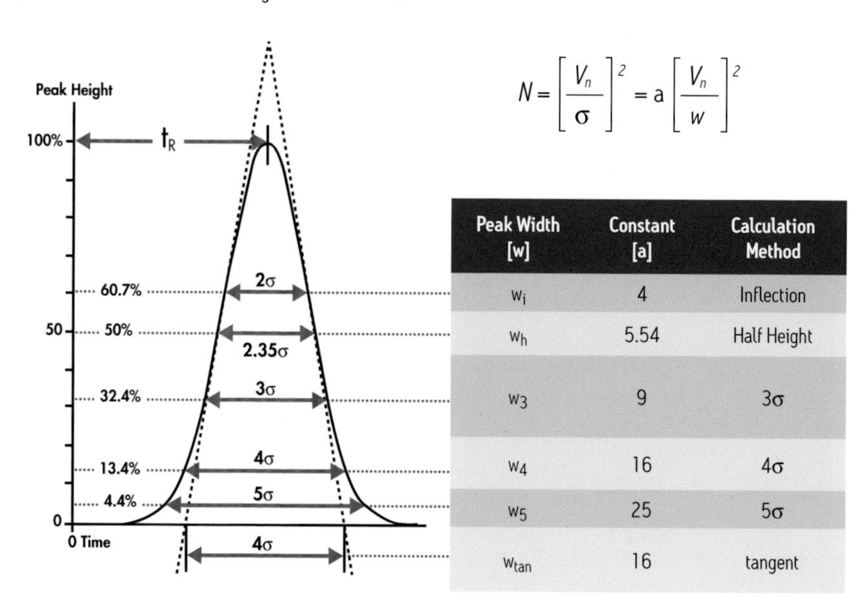

$$N = \left[\frac{V_n}{\sigma}\right]^2 = a\left[\frac{V_n}{w}\right]^2$$

Peak Width [w]	Constant [a]	Calculation Method
w_i	4	Inflection
w_h	5.54	Half Height
w_3	9	3σ
w_4	16	4σ
w_5	25	5σ
w_{tan}	16	tangent

Figure 10: Equation for determining plate count. The narrower the peak width [w], the higher the plate count.

It is a common misconception that plate count refers only to the performance achieved by the column itself. However, the band spreading from both the column and the instrument also influence the peak width from which the plate count is determined. To demonstrate the influence of the instrument itself on band spreading, a single HPLC column was run on two different instruments; a standard HPLC [band spread = 7.2 μL] and an ACQUITY UPLC System [band spread = 2.8 μL]. Since the same column is run on both systems, the band spreading contribution of the column is held constant. The increase in efficiency observed on the ACQUITY UPLC System demonstrates how an instrument with less band spreading will generate narrower peak widths, resulting in a higher plate count [Figure 11].

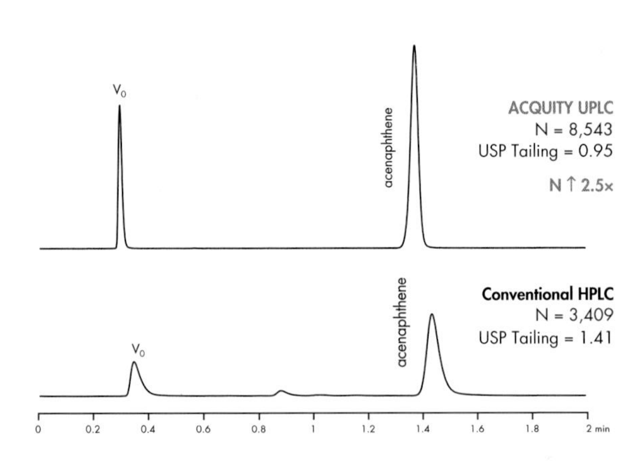

Figure 11: The significant impact of instrument band spreading on column performance. The same column was run on an ACQUITY UPLC System and a conventional HPLC system. [ACQUITY UPLC BEH C_{18} 2.1 x 50 mm, 1.7 µm column; flow rate = 0.4 mL/min.]

The Influence of Tubing Length and Diameter

After the sample band is introduced into the fluidic stream, it travels to the column. The ACQUITY UPLC Sample Manager was designed to minimize the distance between the injector and column inlet in order to minimize band spreading. The column separates the sample band into individual analyte bands. Then the chromatographically separated analyte bands are transferred from the column to the detector. At first glance, one may not consider the importance of the internal diameter [ID] of the tubing connecting the column outlet and detector inlet. However, the internal diameter can greatly impact instrument band spreading [Figure 12].

As one may expect, band spreading will decrease with decreasing tubing ID. By flowing a concentrated analyte band into a large ID tube, the band will become much less concentrated resulting in a wider, distorted peak and decreased sensitivity. Additionally, friction along the walls of the tubing will cause analyte molecules in contact with the tubing walls to move slower than molecules in the center of the tube, resulting in a distorted band profile. The distance between the center and the outer edges of the analyte band become less as tubing ID decreases. This reduces the amount of distortion in the band. It is equally important to minimize the length of tubing, as excessive lengths can also distort a sample band.

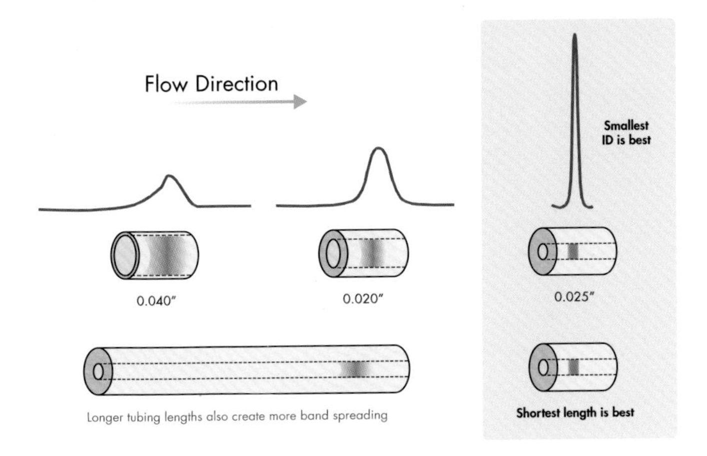

Figure 12: Analyte band spreading as a function of tubing ID and length.

The Impact of Detector Settings on Extra-column Band Spreading

In addition to fluidic-based band spreading contributions of the instrument, digital detector settings related to acquisition speed and filter constant also influence the chromatographic result. This is especially important for UPLC applications, since peak widths are often very narrow [1–2 seconds wide], and analysis time can be very short. When setting the acquisition speed of the detector, the selection should be based upon acquiring sufficient data points across a peak to properly represent the peak. Setting the detector rate too high will negatively impact the signal-to-noise ratio causing the baseline noise to increase without an increase in the signal height of the analyte of interest. Conversely, setting the detector acquisition rate too slow will result in inadequate data points across the peak, thus reducing the observed chromatographic efficiency and compromising the ability to reproducibly perform quantitation. In addition, a time constant [digital filter] is used to smooth out data points to optimize signal-to-noise and can be used conjointly with or independent from, the acquisition speed.

Detector settings become increasingly more important as the width of the analyte band narrows. This requires an acquisition speed that is not only fast enough to digitally create a peak of that speed, but also to properly render the high resolution separation achieved in the UPLC column, if closely eluting peaks exist.

At slower flow rates when peaks are more disperse in time, these settings are more forgiving. However, as the peaks become narrower and the analysis more rapid [as in UPLC separations], a judicious setting of acquisition speed and time constant must be maintained [Figure 13]. When using UPLC Technology for 'real-life' applications, it is wise to select the detector data acquisition rate [Hz] that accurately captures the peak shape of the narrowest peak, and then apply a filter time constant [seconds] that gives optimal signal-to-noise and resolution for the analysis.

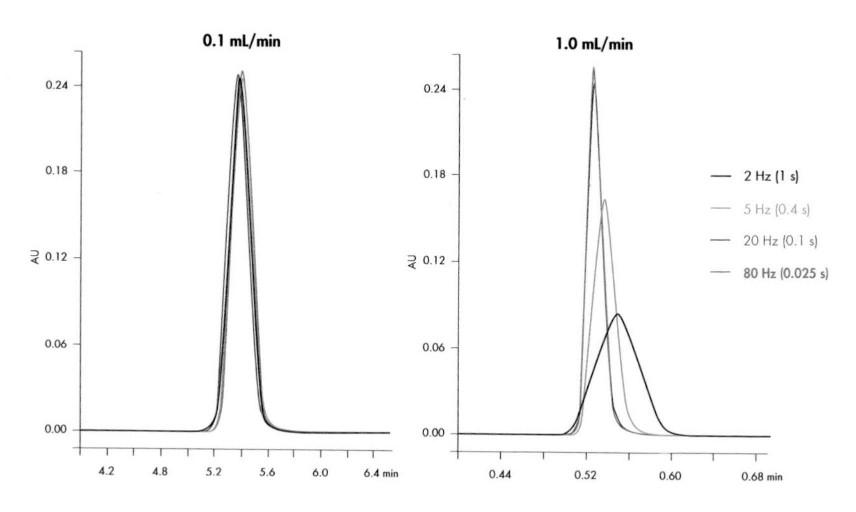

Figure 13: Impact of acquisition speed [Hz] and time constant[s] on peak formation. UPLC peak widths can be very small [1–2 seconds wide]. Therefore, it is critical that proper detector settings are used. Probe is acenapthene analyzed isocratically on an ACQUITY UPLC BEH C_{18} 2.1 x 50 mm, 1.7 µm column; mobile phase 65/35 ACN/H_2O; ACQUITY UPLC System.

The Promise of Small Particles

The Relationship between Resolution and Intra-column Band Spreading

If we think about chromatographic resolution in the most basic sense, it is simply the width [w] of two peaks relative to the distance [$t_{R,2} - t_{R,1}$] between those peaks. If we can make those peaks narrower or further apart, we can improve resolution.

$$Rs \equiv \frac{t_{R,2} - t_{R,1}}{\frac{1}{2}(w_1 + w_2)} = \underbrace{\frac{\sqrt{N}}{4}}_{\text{Efficiency}} \underbrace{\left(\frac{\alpha - 1}{\alpha}\right)}_{\text{Selectivity}} \underbrace{\left(\frac{k}{k+1}\right)}_{\text{Retentivity}}$$

Figure 14: Fundamental resolution equation. [N] is plate count, [α] is selectivity and [k] is retention factor.

Resolution can be expressed mathematically in more relevant terms in the form of the fundamental resolution equation. The resolution equation is comprised of physical and chemical parameters that affect chromatographic resolution; efficiency [N], selectivity [α] and retentivity [k]. Selectivity and retentivity are chemical factors that have historically been easier to manipulate to improve resolution. These parameters can be affected by changes such as temperature, elution solvent, mobile-phase composition and column chemistry. Efficiency is a physical [mechanical] parameter that is more difficult to manipulate due to its square root influence on resolution. However, efficiency can have a significant impact on resolution if the particle size is very small. UPLC Technology focuses on improving resolution by utilizing sub-2 µm particles to improve system efficiency.

It is easier to understand how these parameters affect resolution if we think about how these three terms are represented chromatographically [Figure 15]. Retentivity [k] and selectivity [α] are chemical factors that move peaks relative to one another and are a measure of the interaction of the analytes with the stationary phase and the mobile phase. We can improve resolution by increasing k. However, longer retention times, lower sensitivity and wider peak widths result. An increase in α can result in more resolution, the same peak elution order in a similar amount of time, and/or an elution order change. Efficiency [N] is a physical measure of band spreading in a separation. Assuming that N is improved by reducing particle size of the packing material, the center-to-center peak distance does not change. Additionally, a reduction in particle size will result in narrower, more efficient chromatographic peaks, thus improving resolution and sensitivity.

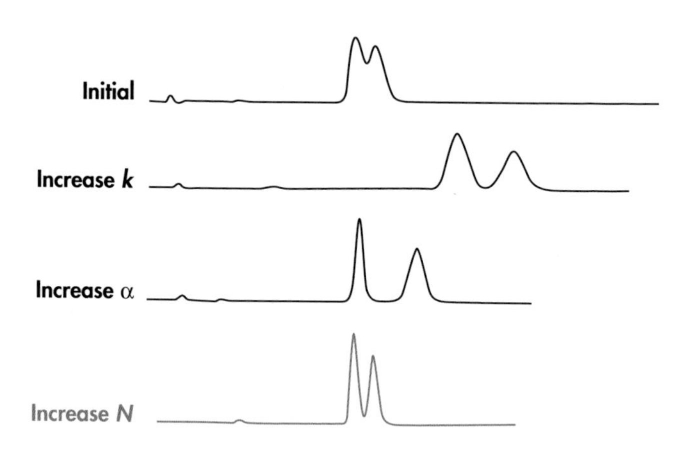

Figure 15: The impact of individual chemical and mechanical factors on resolution.

The Relationship between Resolution, Efficiency and Particle Size

UPLC Technology maximizes the physical [mechanical] contribution to resolution by minimizing instrument band spreading, enabling the use of higher efficiency, smaller particle columns [1.7 µm – 1.8 µm]. With simple chromatographic examples and basic arithmetic, one can gain a better understanding of the chromatographic principles behind UPLC Technology.

As stated in the fundamental resolution equation, resolution is directly proportional to the square root of efficiency.

$$Rs \propto \sqrt{N}$$

Figure 16: Resolution [Rs] is directly proportional to the square root of efficiency [N].

Additionally, efficiency is inversely proportional to the particle size. This means, if the particle size of the packing material is decreased, separation efficiency increases. For example, if the particle size of the packing material is reduced from 5 µm to 1.7 µm [3×], theory predicts that efficiency should increase 3×, resulting in a 1.7× increase in resolution [square root of 3].

$$N \propto \frac{1}{d_p} \qquad d_p \downarrow 3\times \quad N \uparrow 3\times \quad Rs \uparrow 1.7\times$$

Figure 17: At constant column length, efficiency [N] is inversely proportional to particle size [d_p].

In order to achieve the gains of efficiency and resolution predicted by theory, the optimal flow rate must be run with respect to the particle size. The optimal flow rate [F_{opt}] is inversely proportional to the particle size. This means, if the particle size is reduced from 5 µm to 1.7 µm [3×], the optimal flow rate in which to run that particle increases 3×, resulting in a reduction in analysis time to the same degree [3×], therefore increasing sample throughput.

$$F_{opt} \propto \frac{1}{d_p} \qquad d_p \downarrow 3\times \quad Rs \uparrow 1.7\times \quad T \downarrow 3\times$$

Figure 18: At constant column length, flow rate [F_{opt}] is inversely proportional to particle size [d_p], resulting in a reduction of analysis time [T] proportional to the reduction in particle size.

One of the first things we observe as chromatographers is what happens when the chromatographic flow rate is changed. If the flow rate is increased, analytical run time decreases. Additionally, the width of the peak also decreases. As peak width becomes narrower, the height of that peak increases proportionally. Narrower peaks that are taller, are easier to detect and differentiate from baseline noise [higher S/N], resulting in higher sensitivity.

$$N \propto \frac{1}{w^2} \qquad N \uparrow 3\times \quad w \downarrow 1.7\times \quad \text{peak height} \uparrow 1.7\times$$

Figure 19: Narrower peak widths produce higher efficiency and peak heights. Efficiency [N] is inversely proportional to the peak width [w] squared. Further, as the peak width decreases, the peak height increases proportionally.

These theoretical principles were applied chromatographically. As observed in Figure 20, when extra-column band spreading is minimized as in the ACQUITY UPLC Instrument, the theoretical performance of a column can be achieved.

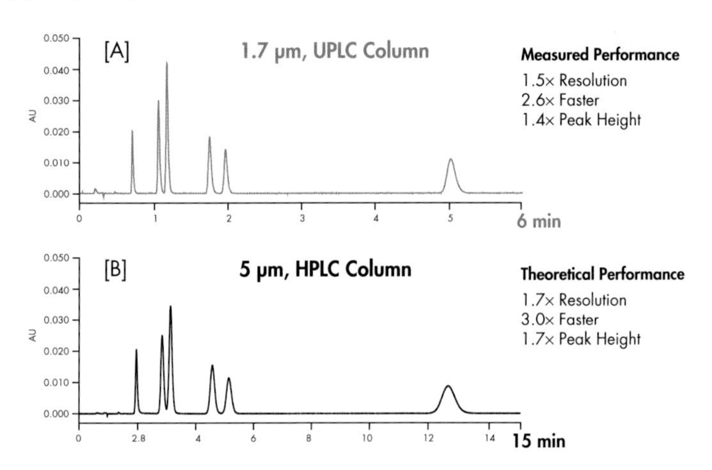

Figure 20: Matching theory to reality. Separations were performed on two columns with same dimensions [2.1 x 50 mm]. Identical chromatographic conditions were used in both separations with the exception of flow rate, which were scaled based on particle size.

The discussion above demonstrates the importance of intra-column band spreading. If one can further understand what processes influence band spreading and how to reduce it, improvements in efficiency, and therefore resolution, can be achieved.

Understanding van Deemter Curves

As described earlier, the width of a peak can be thought of as a statistical distribution of the analyte molecules [variance, σ^2]. The peak width increases linearly in proportion to the distance in which that peak has traveled. The relationship between peak width and the distance in which that peak has traveled, is a concept called the height equivalent to a theoretical plate [HETP or H]. Originating from distillation theory, H is a measurement of column performance that takes into account several band spreading related processes. To put this into terms that may be more familiar, the smaller the HETP, the more plates [N] there are in a column [Figure 21].

$$\text{HETP} = \frac{L}{N}$$

Figure 21: Simplified equation to determine HETP. [L] is column length, [N] is plate count and [HETP] is height equivalent to a theoretical plate.

If we think about what is occurring at a molecular level within a column [how the analyte molecules are interacting with the mobile phase and stationary phase], we can further understand the different diffusion-related processes that are occurring that contribute to chromatographic performance [Figure 22].

There are several diffusion-related processes occurring simultaneously:

1) Analyte molecules are transported to the surface of the particle and around that particle [eddy diffusion]

2) Analyte molecules are diffusing back and forth in the bulk mobile phase [longitudinal diffusion]

3) Analyte molecules are diffusing into and out of the chromatographic pores [mass transfer].

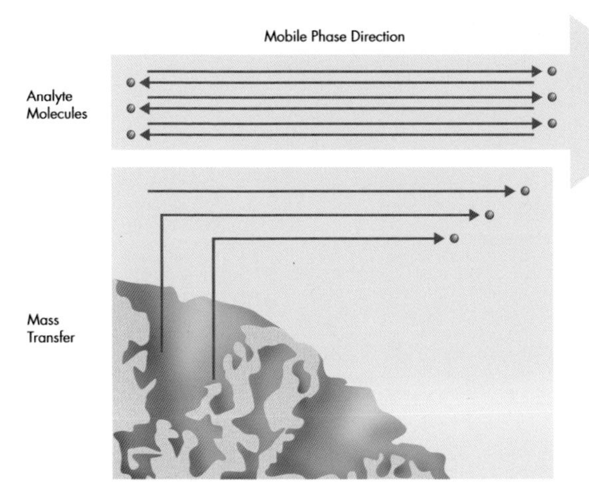

Figure 22: Diffusion-related processes occurring within the column.

These diffusion-related processes can be expressed mathematically in the form of the van Deemter equation.

$$\mathrm{HETP} = a\,(d_p) + \frac{b}{u} + c\,(d_p)^2 u$$

A term B term C term

Figure 23: van Deemter equation.

The van Deemter equation is comprised of three terms:

The A term [eddy diffusion] is related primarily to the particle size of the packing material. Its value is also determined by how well the chromatographic bed is packed. It is also related to the uniformity or non-uniformity of the flow to and around a particle.

The B term [longitudinal diffusion] is related to the diffusion of the analyte in the bulk mobile phase and on the stationary phase and decreases with increasing speed of the mobile phase [linear velocity].

The C term [mass transfer] is related to both the linear velocity [the speed of the mobile phase] and the square of the particle size. Mass transfer is the interaction of analyte molecules with the internal surface of the stationary phase and their distance of diffusion into and out of the pores of the packing material.

We can observe these terms individually by plotting them on a scale of HETP vs. the linear velocity [u] of the mobile phase [Figure 24].

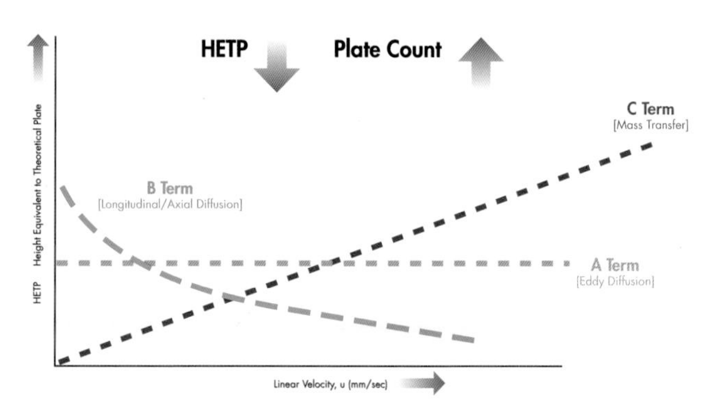

Figure 24: Individually plotted terms of the van Deemter equation.

The A term is plotted as a horizontal line. It is related to the particle size and how well a column is packed and is independent of linear velocity [mobile-phase speed]. As the particle size of the packing material is decreased, the H value also decreases [higher efficiency].

The B term is plotted as a downward sloping curve with increasing linear velocity. This term is independent of the particle size and indicates that if the mobile phase moves at a slower linear velocity, analyte molecules reside in the column for a longer time, and therefore, a greater opportunity exists for band spreading [diffusion] lengthwise within the column. Conversely, if the mobile phase moves at a faster linear velocity, there is less time for diffusion, and therefore there is less time for band spreading to occur.

The C term is plotted as an increasing linear relationship between H and u. Among a distribution of molecules, some molecules enter a stationary-phase pore while others remain in the moving mobile phase until they reach another particle. This is followed by the reverse process, where immobilized molecules detach and move further down the chromatographic bed. It takes time for a molecule to move into and out of the pores, and therefore, as molecules transfer from one particle to the next, the analyte band that contains those molecules broadens as it moves down the length of the column. The smaller the stationary phase particles, the faster this process occurs keeping the analyte band from spreading. If the mobile phase moves fast, a larger distance develops between the immobilized molecules and the ones that move ahead. This indicates that in order for the population of analyte molecules to remain together, the mobile phase should move at a slower linear velocity. As the speed of the mobile phase is increased, the population of analyte molecules will become more disperse, resulting in increased band spreading.

A van Deemter curve is formed upon adding the A, B and C terms together [Figure 25]. Conducting a separation at the linear velocity at which the lowest point in the curve occurs will yield the highest efficiency and therefore, chromatographic resolution.

If we correlate this to the van Deemter equation shown in Figure 23, if particle size is reduced by half, H is reduced by a factor of 2. Hence, it is possible to reduce band spreading within a column by utilizing smaller particles.

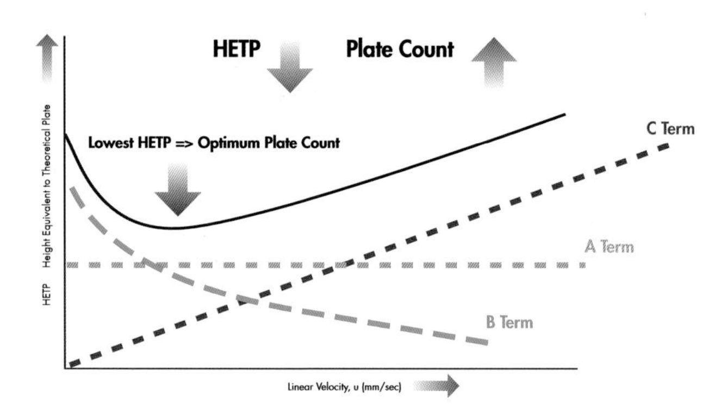

Figure 25: Adding the three individual terms of the van Deemter equation together yields a van Deemter curve.

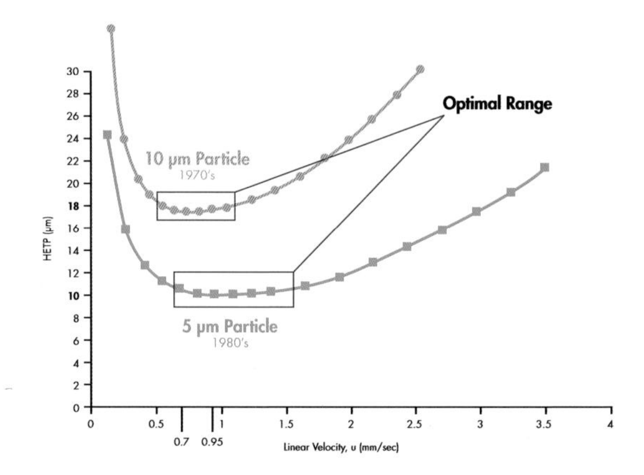

Figure 26: Van Deemter plot comparing 10 µm and 5 µm particles.

For example, Figure 26 depicts a van Deemter plot of a 10 µm particle and a 5 µm particle. As observed on the plot, a large 10 µm particle has a very narrow optimal operating range with respect to linear velocity, to achieve the lowest H value [18 µm @ 0.7 mm/sec]. If the speed of the mobile phase is too slow or too fast, an increase in H [loss of efficiency] is observed, thus reducing chromatographic resolution and sensitivity. However, the 5 µm particle exhibits a much lower H value, [10 µm @ 0.95 mm/sec; higher efficiency] at a higher mobile-phase speed, as well as a larger linear velocity range in which that H value can be achieved. This means, a higher efficiency, and therefore resolution, can be achieved in a faster time than with a column packed with larger 10 µm particles.

To gain additional understanding of this effect, we can further investigate the influence of particle size on the individual terms of the van Deemter equation.

Particle size has a significant impact on the analyte band as it relates to the A term [eddy diffusion]. The path which analyte molecules take to transfer from the bulk mobile phase to the surface of the particle and around that particle takes less time as particle size is decreased. Larger particles cause analyte molecules to travel longer, more indirect paths. The differences in these paths result in different migration times for the analyte molecules within a population, resulting in a broader analyte band and resulting peak. As the particle size of the packing is decreased, the paths of the analyte molecules are encouraged to be more similar in length. This results in narrower analyte bands which translate into narrower peaks, higher efficiency and higher sensitivity [Figure 27].

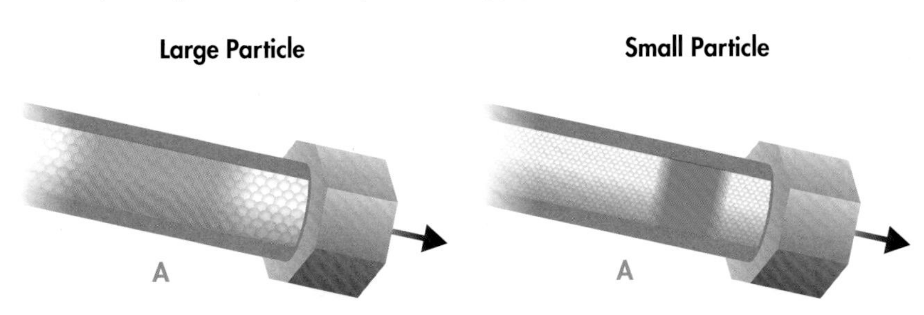

Figure 27: The influence of particle size on the A Term.

Longitudinal diffusion, the B term, is not directly impacted by particle size. However, as particle size decreases, the other parameters of the van Deemter equation, the A and C term, become smaller. Consequently, the optimal linear velocity [mobile-phase speed] increases, resulting in less opportunity for band spreading. Slower linear velocity allows the band of analyte molecules to interact with the packing material for a longer period of time. This means there is more time for axial [lengthwise] diffusion into the mobile phase, resulting in broader, more diffuse analyte bands. At higher linear velocity, the population of analyte molecules is swept through the column in a shorter period of time which enables the analyte band to remain more concentrated, resulting in narrower, higher efficiency peaks due to less time for longitudinal diffusion [Figure 28].

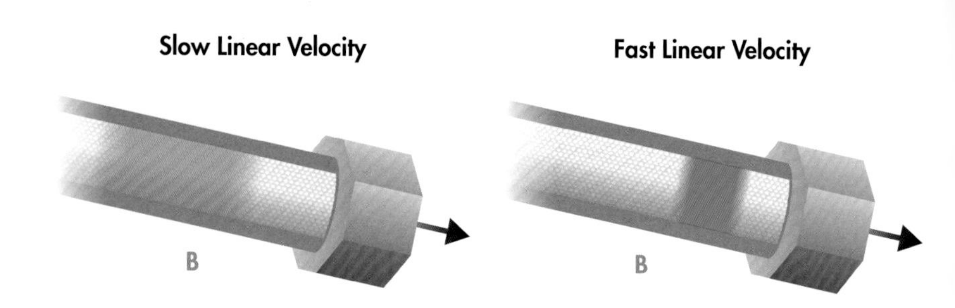

Figure 28: Influence of linear velocity on longitudinal diffusion, the B term.

The C term [mass transfer] is impacted by both linear velocity and particle size. The populations of analyte molecules are transported from the mobile phase to the particle surface. The analyte molecules then move through the mobile phase in the pores to the bonded surface layer (e.g. C_{18}, C_8, etc.), interact with the bonded phase and then are swept back out of the pore into the bulk mobile phase. However, the analyte molecules within the population travel into and out of the pore to varying degrees. That means as the molecules return to the bulk mobile phase, the length of the path in which each analyte molecule has traveled is different resulting in a spreading [widening] of the analyte band [Figure 29]. The amount of band spreading that occurs will depend on the speed of the mobile phase [Figure 30].

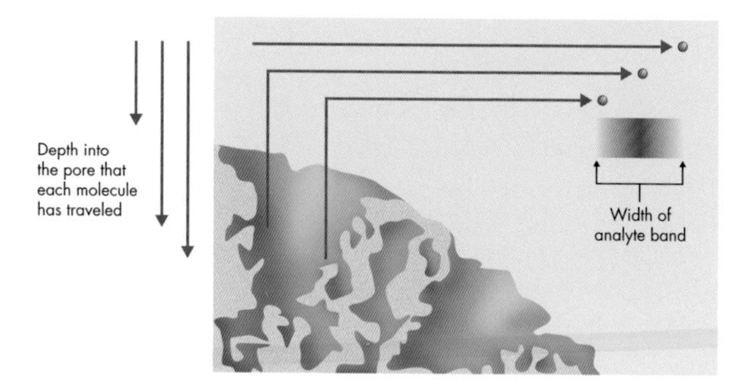

Figure 29: Mass transfer [diffusion] into and out of a chromatographic pore.

High Linear Velocity

Large lag time for analytes travelling into pores when linear velocity is high

Slow Linear Velocity

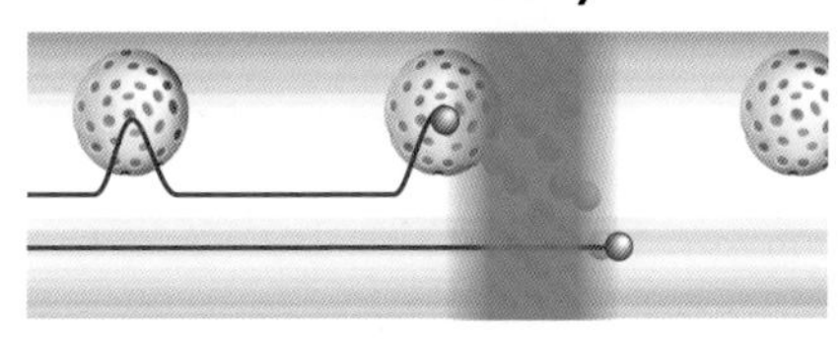

Narrower, more concentrated band resulting in sharper peaks

Figure 30: Impact of linear velocity on mass transfer and analyte bands [same particle size].

At a high linear velocity, the time between the molecules interacting with the particle surface and transferring through the mobile phase is longer. The faster the speed of the mobile phase, the quicker the analyte molecules will move through the column, resulting in a broader, less concentrated analyte band. This translates into a broader chromatographic peak and lower sensitivity.

At a slow linear velocity, the length of the steps between interactions to the surface is shorter. This results in a more concentrated analyte band, producing narrower, more efficient chromatographic peaks.

Mass transfer improves dramatically as the particle size is decreased due to its relationship to the square of the particle size [d_p^2]. For smaller particles, it takes less time for an analyte molecule to travel into the pores, interact with the chromatographic surface, and be swept back into the mobile phase. Therefore, analyte molecules separated on smaller particle columns diffuse much faster, resulting in a sharper, narrower and more efficient chromatographic band [Figure 31].

1.7 µm Particle Pore

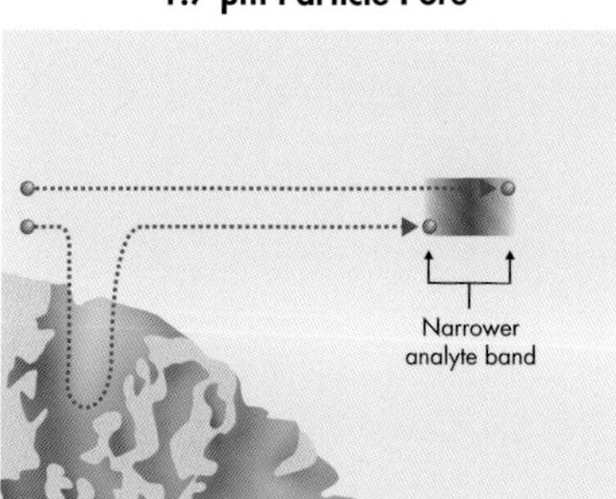

5 µm Particle Pore

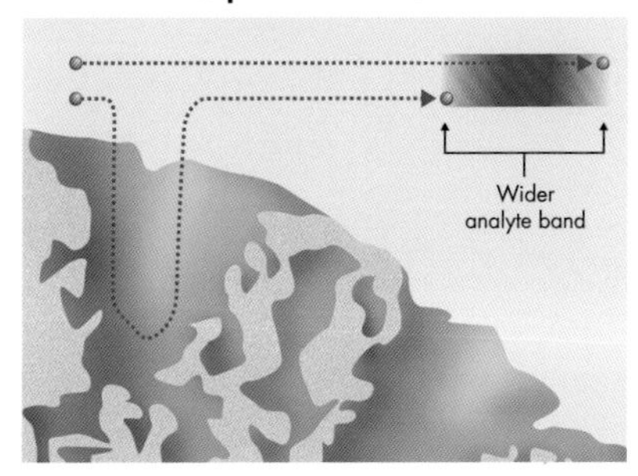

Figure 31: Mass transfer differences related to particle size [representation of a 100Å pore]. Narrower analyte bands are formed with smaller particles.

Reducing particle size will therefore improve mass transfer, thus effectively decreasing the slope of the C term, which leads us to where we are today with UPLC Technology. As can be seen in the van Deemter plot in Figure 32, 1.7 µm particles provide 2-3× lower HETP values than 3.5 µm particles. Additionally, these lower H values are achieved at a higher linear velocity and over a broader range of velocities. This means that mass transfer is improved dramatically with the smaller particle, enabling better efficiency and resolution. It also means that one can use an increased range of linear velocities to gain this improved performance. Separations are able to be performed at faster linear velocities, thus improving speed of analysis, without compromising resolution.

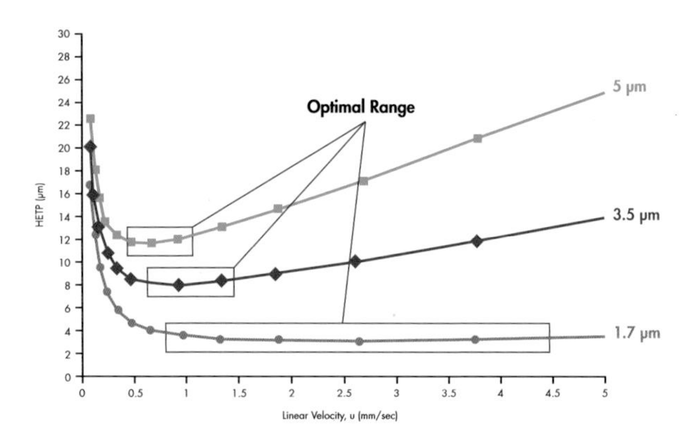

Figure 32: van Deemter plot comparing particle size.

The Impact of Extra-column [Instrument] Band Spreading on van Deemter Plots

As UPLC Technology has gained momentum and popularity in the chromatographic laboratory, van Deemter plots have emerged as a popular method to measure the performance gains of sub-2 µm particles relative to existing HPLC particle sizes. Erroneous conclusions can be made if these comparative measurements are not carried out on instrumentation that minimizes the contribution of extra-column band spreading.

To demonstrate the importance of extra-column band spreading on performance measurement, van Deemter plots were generated on a conventional HPLC instrument [band spread = 7.2 µL], comparing 1.7 µm vs. 2.5 µm particles [Figure 33]. The particle substrate and bonded phase chemistry of the two columns were identical. At first glance, the van Deemter plots infer there is no appreciable difference in the performance of these two columns. How can this be?

In this case, the band spreading of the HPLC instrument results in a similar performance measurement of the 1.7 µm particle UPLC column relative to the 2.5 µm HPLC column. The smaller peak widths produced by 1.7 µm particle UPLC columns are impacted more greatly by extra-column band broadening than columns packed with larger particles [e.g., 2.5 µm], therefore producing this misleading result.

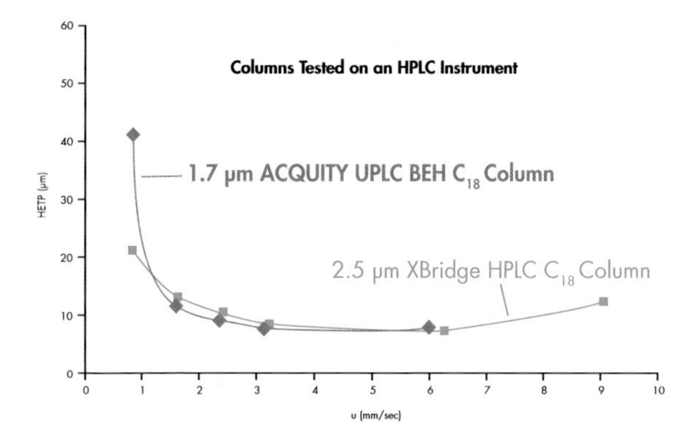

Figure 33: Sub-3 µm particle comparison on an HPLC instrument resulting in similar performance and linear velocity range. van Deemter curves for acenaphthene on an XBridge™ HPLC C_{18} 2.1 x 50 mm, 2.5 µm column and an ACQUITY UPLC BEH C_{18} 2.1 x 50 mm, 1.7 µm column.

The same experiment was then conducted on an ACQUITY UPLC Instrument [band spread = 2.8 µL]. The ACQUITY UPLC Instrument has approximately 84% lower system volume and 60% lower band spreading than the HPLC instrument.

As can be seen in Figure 34, a noticeable difference is observed in the performance of these two columns when operated on the ACQUITY UPLC Instrument. In addition to the observed differences in HETP, the optimal linear velocity increases from 3.0 mm/sec [2.5 µm particle] to 10.0 mm/sec [1.7 µm particle], demonstrating the performance gains of lower HETP [higher efficiency] and faster linear velocity [and therefore throughput] associated with UPLC Technology.

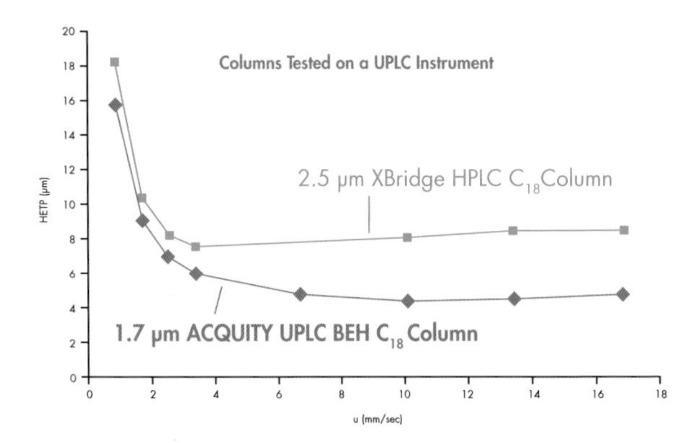

Figure 34: Sub-3 µm particle comparison on an ACQUITY UPLC Instrument resulting in improved performance and linear velocity range with decreasing particle size. van Deemter curves for acenaphthene on an XBridge HPLC C_{18} 2.1 x 50 mm, 2.5 µm column and an ACQUITY UPLC BEH C_{18} 2.1 x 50 mm, 1.7 µm column.

Maintaining Separation Integrity without Compromise

Now that a basic understanding of the functions of extra-column and intra-column band spreading has been established, we can relate the measure of those terms [HETP vs. linear velocity] in a manner that is more practical: plate count vs. flow rate.

As discussed previously, linear velocity is the speed of the mobile phase as it flows through the column. It is a term used to normalize the flow rate of the mobile phase independent of column ID, such that the performance of columns with different dimensions can be measured and compared. The optimal linear velocity is directly related to the optimal flow rate [Figure 35]. Conventional HPLC columns can only be operated in a narrow flow-rate range to achieve the optimum performance. If one operates outside of this range, a decrease in performance can be expected.

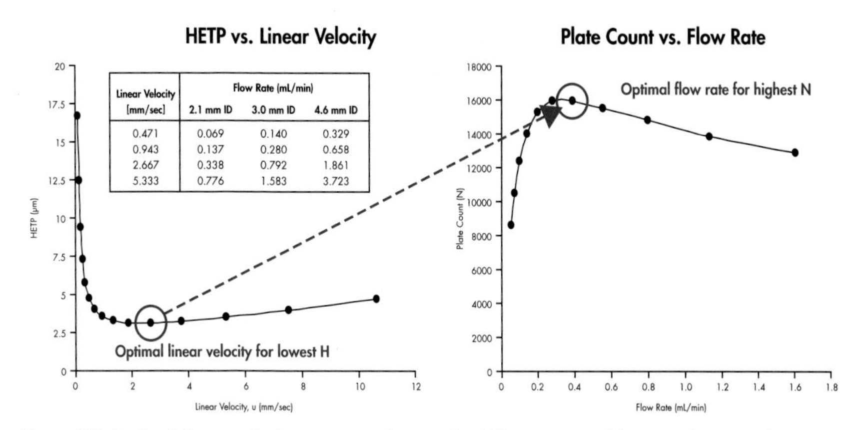

Linear Velocity [mm/sec]	Flow Rate (mL/min)		
	2.1 mm ID	3.0 mm ID	4.6 mm ID
0.471	0.069	0.140	0.329
0.943	0.137	0.280	0.658
2.667	0.338	0.792	1.861
5.333	0.776	1.583	3.723

Figure 35: Optimal linear velocity corresponds to optimal flow rate to achieve maximum performance. Values were calculated for a 2.1 mm ID x 50 mm long column packed with 1.7 µm particles.

A common approach to reduce analysis time [improve throughput] in conventional HPLC is to simply increase the flow rate. With larger particles, increasing the flow rate often results in a substantial loss in efficiency, and therefore resolution, since you are now operating above the optimal linear velocity for the HPLC particle [compressed chromatography]. This is a significant compromise between chromatographic performance and analysis speed that is associated with HPLC [Figure 36].

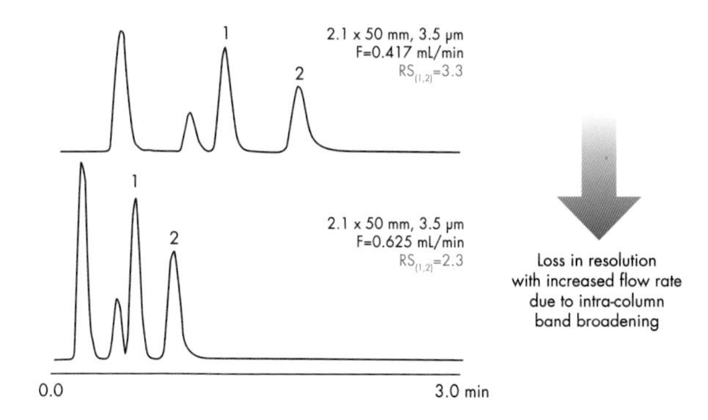

Figure 36: In HPLC, a compromise between resolution and speed must be made. In this case, a 30% loss in resolution is observed.

UPLC Technology is not prone to these compromises. Analysis time can be decreased without sacrificing performance. When particle size is reduced from 3.5 µm to 1.7 µm, a meaningful increase in efficiency is observed [Figure 37]. This is due to the narrow chromatographic bands that are produced by 1.7 µm particle UPLC columns as a result of negligible intra-column band spreading. Additionally, this added efficiency occurs at a higher flow rate [Figure 37]. This means that for the same column length, a dramatic increase in efficiency, and therefore resolution, can be achieved in a shorter analysis time.

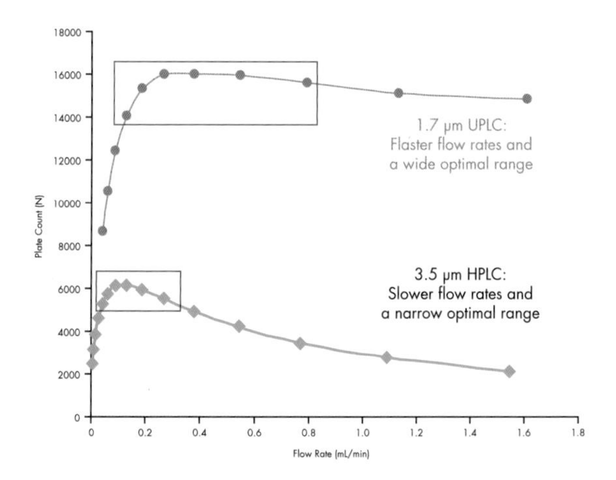

Figure 37: Dependence of particle size on optimal flow rate.

Understanding Column Resolving Power [L/d_p]

When performing a chromatographic separation, the primary goal is to resolve one component from another so that some or all of the components can be measured. The maximum resolving power of a column can be estimated by dividing the column length [L] by the particle size [d_p]. The L/d_p ratio is particularly useful when trying to determine which particle size packing material and column length may be necessary for a given application [Figure 38].

Typical HPLC Column 4.6 x 150 mm, 5 µm

Column Length (L) = 150 mm = 150,000 µm

d_p = 5 µm

$$\frac{L}{d_p} = \frac{150{,}000}{5} = 30{,}000$$

Separation Index	Application Example	Efficiency (N)	L/d_p
Easy	Content uniformity	5,000	15,000
Moderately Challenging	Related compound assay	12,000	30,000
Difficult	Impurity profiling	20,000	50,000
Extremely Difficult	Metabolite identification	35,000	85,000

Figure 38: Calculating the L/d_p ratio.

This ratio can also be used as a tool for transferring methods from one particle size to another. A column that has an L/d_p ratio of 30,000 [moderately challenging separation] is a very common selection. As can be seen in Figure 39, a typical HPLC column that produces a resolving power of 30,000 is 150 mm long and is packed with 5 µm particles. As particle size is decreased, the same resolving power can be achieved in a shorter column [which means faster analysis time; i.e. a 50 mm long column packed with 1.7 µm particles achieves an L/d_p ratio of 30,000]. In addition to shorter column length, the optimal flow rate increases as particle size decreases, which further adds to the reduction in analysis time.

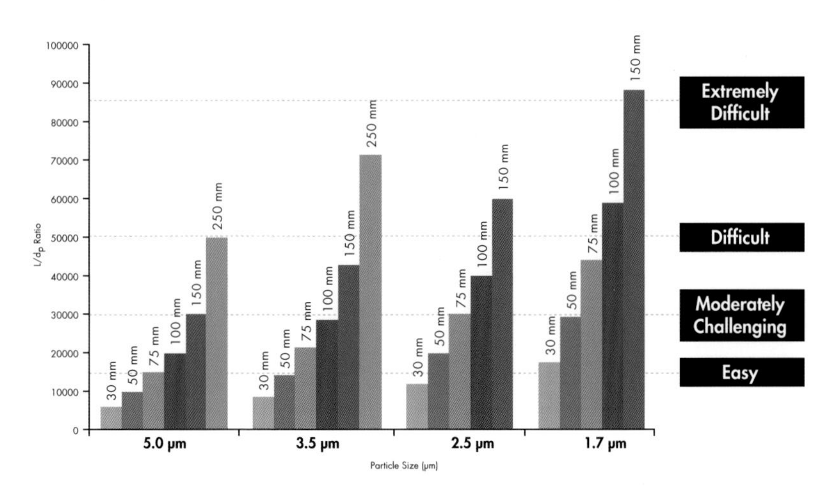

Figure 39: Comparing the L/d_p ratio as a function of separation index [easy-to-extremely difficult]. Columns with the same L/d_p ratio will generate the same resolving power.

This is more clearly demonstrated chromatographically [Figure 40]. A 50 mm long UPLC column packed with 1.7 µm particles produces the same resolving power as a 150 mm long HPLC column packed with 5 µm particles. By keeping the L/d_p ratio constant, analysis time is shortened 10× while resolution was maintained. Flow rates were adjusted in inverse proportion to each particle size. Injection volumes were scaled in proportion to the column volume such that the same mass load on-column was injected.

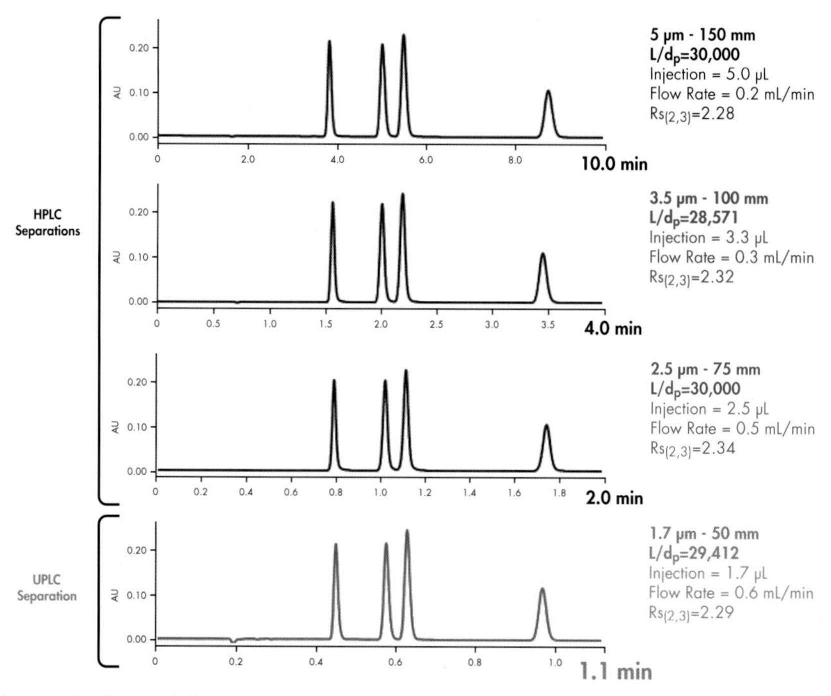

Figure 40: Holding L/d_p constant while reducing particle size enables faster separations while maintaining separation integrity.

Understanding the role of column length to particle size ratio [L/d_p] is key to the understanding of UPLC Technology. UPLC Technology is based upon efficiently packing small, pressure-tolerant particles into short [high-throughput] or long [high-resolution] columns. These UPLC columns are used in an LC instrument designed to operate at the optimal linear velocity [and resulting pressure] for these particles with minimal band spreading.

Measuring Gradient Separation Performance [Peak Capacity]

Under isocratic conditions, plate count [N] is a measure of the cumulative band spreading contributions of the instrument and the column. Due to diffusion related band broadening, the width of an analyte band increases the longer the analyte band is retained on the stationary phase.

In a gradient run, the elution strength of the mobile phase changes over the course of the analysis. This causes stronger retained analyte bands to move more quickly through the column [thus changing retention time], keeping the bands more concentrated [narrow]. In reversed-phase chromatography, the increasing elution strength of the mobile phase controls the width of the bands being produced, resulting in similar peak widths as the bands pass through the detector. Since peak width and retention time are being altered by the changing strength of mobile phase, plate count [due to its relationship to peak width] is not a valid measurement for gradient separations.

The resolving [separation] power of a gradient can be calculated by its peak capacity [P_c]. Thus, peak capacity is simply the theoretical number of peaks that can be separated in a given gradient time. Peak capacity is inversely proportional to peak width. Therefore, for P_c to increase, peak width must decrease.

$$P_c = 1 + \frac{t_g}{w}$$

Figure 41: Peak capacity [P_c] equation, where [t_g] is the gradient time and [w] is the average peak width.

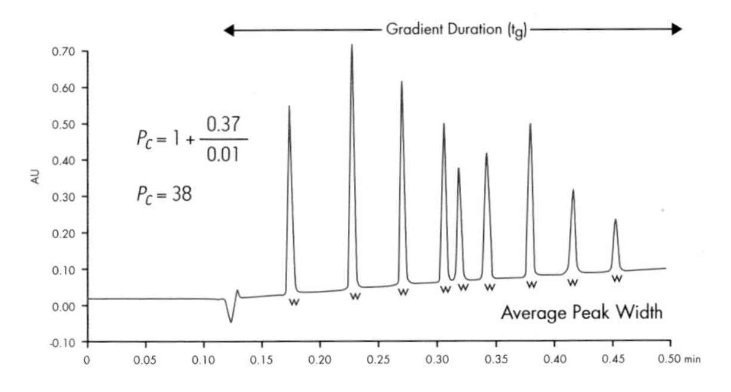

Figure 42: Applying the peak capacity equation to a fast separation, where 0.37 minutes is the gradient duration time and 0.01 minutes is the average peak width, resulting in a peak capacity of 38. Peak width was measured at 13.4% peak height [4σ].

Peak capacity can be dramatically increased by using UPLC Technology. The high resolving power of UPLC Technology enables more information per unit time to be generated by harnessing the power of sub-2 µm particles on exceptionally low dispersion instrumentation [2.8 µL band spread]. This facilitates more information that can be collected from a given sample. For example, a tryptic digest of phosphorylase b yields ~70 identified peaks using an HPLC column packed with 5 µm particles [Figure 43A]. With UPLC Technology, the number of identifiable peaks increases from 70 to 168, thus improving confidence in protein identification [Figure 43B].

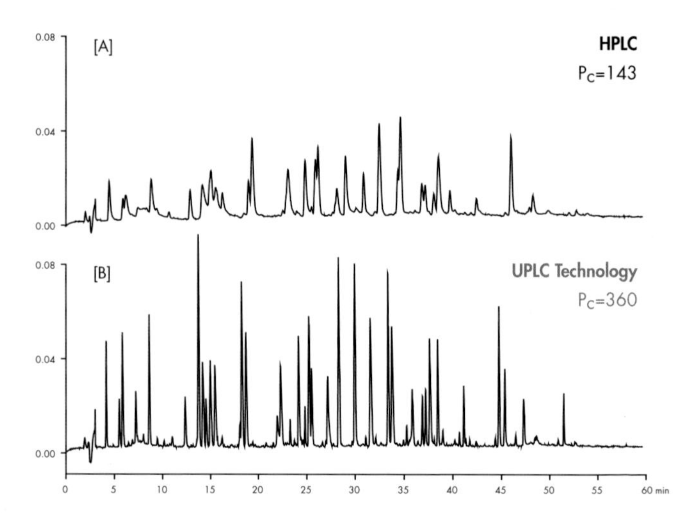

Figure 43: Peak capacity comparison of HPLC vs. UPLC Technology.

The Consequence of Improved Performance

Addressing System Pressure

In the previous section, the benefit of smaller chromatographic particles and minimized system band spreading [both instrument and column] was established. UPLC Technology facilitates improved chromatographic performance by minimizing system band spreading to produce more efficient separations in less time, thus achieving better data quality. Band spreading, however, is not the only factor that dictates the performance one can achieve with small particles. The available pressure of the instrument also plays a large role.

Pressure is inherently generated when mobile phase passes through the connective tubing from the pump to the injector, the injector to the column, the column itself, the tubing post column as well as the detector cell. The measurement of system pressure is a cumulative effect of all of these components [instrument and column]. As flow rate is increased, the pressure produced by mobile phase flowing through the connective tubing itself will increase. Additionally, the ID of the tubing, as well as its length, will also impact how much pressure will be generated in combination with the flow rate. The pressure difference between two columns can be compared against theoretical predictions if the pressure generated by the instrument itself, is subtracted from the total system pressure [instrument + column].

As particle size is reduced, back pressure increases at a rate that is inversely proportional to the square of the particle diameter. Simultaneously, the optimal mobile-phase speed [linear velocity] increases with decreasing particle diameter. Therefore, the pressure at the optimal linear velocity for a given particle size increases at a rate that is inversely proportional to the cube of the particle diameter [Figure 44].

$$\Delta P_{opt} = \frac{1}{d_p^{3}} \qquad d_p \downarrow 3\times \qquad pressure \uparrow 27\times$$

Figure 44: The relationship between optimal pressure [ΔP_{opt}] and particle size [d_p] for a constant column length. If particle size is decreased by a factor of 3, pressure will increase 27×.

This is a significant limitation when trying to utilize smaller particle columns on conventional HPLC instrumentation to improve chromatographic resolution [keeping column length constant while reducing particle size] or to improve the speed of analysis while maintaining resolution [keeping the L/d_p ratio constant]. Due to pressure limitations of conventional HPLC instruments

[350–400 bar; 5000–6000 psi], the use of smaller particles often results in a restriction of column length or operating at sub-optimal linear velocities [flow rates].

For a constant column length, theory predicts if particle size is decreased from 5.0 µm to 1.7 µm [3× decrease in particle size], backpressure is anticipated to increase 27×. Matching close to theoretical predictions, system pressure increased 22× when transitioning from a 5.0 µm column to a 1.7 µm column of the same length. As observed, the 1.7 µm column is operating well above the upper pressure limit of a conventional HPLC instrument [Figure 45].

The large increase in backpressure observed by decreasing particle sizes is one of the primary reasons why sub-2 µm particle columns [and corresponding LC instruments] were never commercially successful until the advent of the ACQUITY UPLC System.

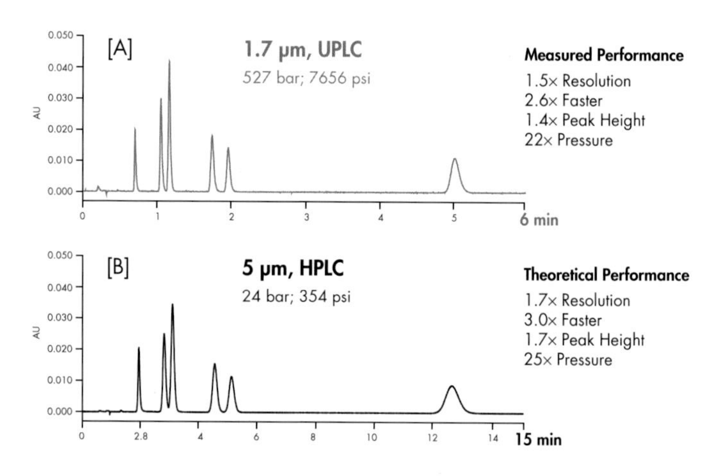

Figure 45: Influence of particle size and optimal flow rate on column pressure [subtracted from total system pressure]. Constant column length. 2.1 x 50 mm columns; flow rate = 0.6 mL/min [1.7 µm] and 0.2 mL/min [5 µm].

If the separation goal is to maintain resolution while reducing analysis time [keeping the L/d_p ratio constant], the increase in pressure is much less than holding column length constant while reducing particle size. The change in pressure is inversely proportional to the square of the particle size [rather than the particle size cubed] due to the proportional reduction in column length.

$$\Delta P_{opt} = \frac{1}{d_p^2} \qquad L \downarrow 3\times \qquad d_p \downarrow 3\times \qquad pressure \uparrow 9\times$$

$$P \propto L$$

Figure 46: The relationship between optimal pressure [ΔP_{opt}] and particle size [d_p] for differing column length. If particle size and column length are decreased by a factor of 3, pressure will increase 9×.

In this example, both column length and particle size are reduced 3× [Figure 47]. This means backpressure is anticipated to increase 9×. The observed values match closely with theoretical predictions. Keeping L/d_p constant, an 11× increase in back pressure is observed when transitioning from a 5.0 µm, 150 mm long column to a 1.7 µm, 50 mm long column.

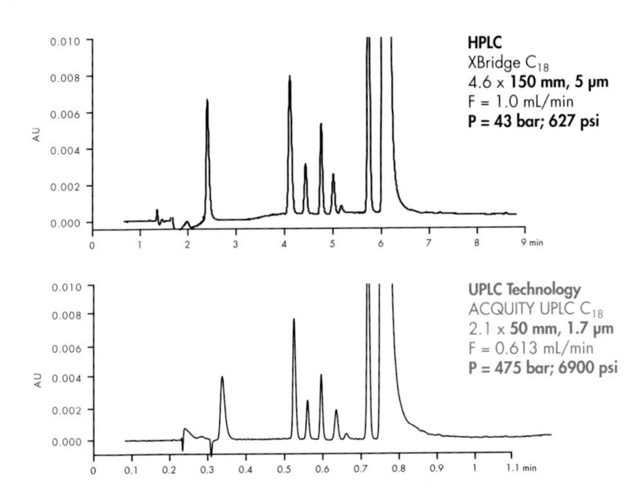

Figure 47: Influence of particle size, column length and optimal flow rate on column pressure [subtracted from total system pressure]. Constant L/d_p ratio. Note the significant difference in analysis time [the UPLC separation is 7× faster].

When run at their optimal flow rate, the pressure produced by smaller particles will exceed the pressure limitations of conventional HPLC systems. The ACQUITY UPLC System [upper pressure limit of 1030 bar, 15,000 psi] was designed to accommodate these pressures, enabling sub-2 µm particles to be successfully run at their optimal flow rate.

Elevated Temperature

One approach towards compensating for the higher pressures produced by small particles is to elevate the column temperature. As column temperature is increased, mobile-phase viscosity decreases, resulting in lower backpressure [if flow rate is held constant]. However, the speed at which analyte molecules move in and out of the stationary phase pores [diffusion] also increases, resulting in the need to increase flow rate in order to maintain performance.

When increasing column temperature from 30 °C to 90 °C, flow rate must be increased in order to maintain efficiency [Figure 48]. No gain in efficiency is observed when comparing the highest plate count at either temperature which is in agreement with chromatographic theory.

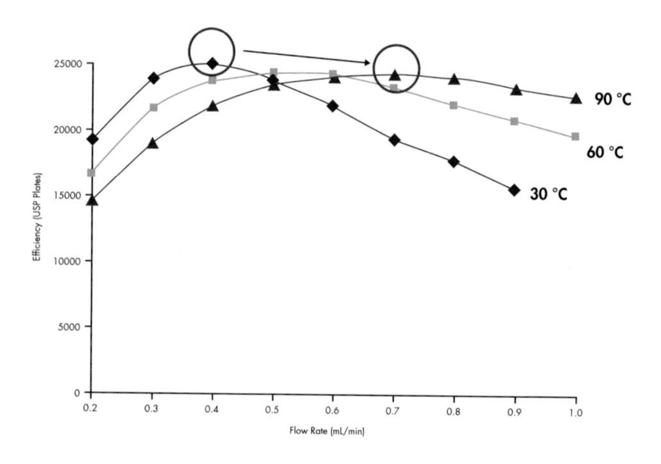

Figure 48: The effect of column temperature on efficiency. Isocratic retention of amylbenzene on an ACQUITY UPLC BEH C_{18} 2.1 x 100 mm, 1.7 µm column.

A more interesting comparison can be made if plate count is plotted against system pressure [Figure 49]. By plotting the data in this way, one can clearly see that the maximum column efficiency is achieved at approximately the same system pressure, independent of the temperature of the separation. This means that elevated temperature cannot be used to circumvent the pressures associated with the use of small particles. In other words, a conventional HPLC instrument is not suitable for the efficient use of very small particles.

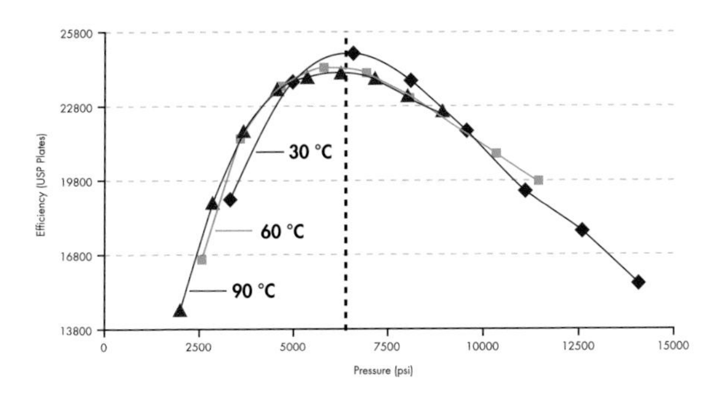

Figure 49: Maximum efficiency is achieved at similar pressures, independent of temperature. Isocratic retention of amylbenzene on an ACQUITY UPLC BEH C_{18} 2.1 x 100 mm, 1.7 µm column.

Improving Productivity with UPLC Technology

Holistic System Design

By understanding the chromatographic principles outlined within this primer and how they are applied, it becomes apparent that more than just small particles and higher pressure must be considered to maximize separation performance. In order to attain the chromatographic benefit from sub-2 µm particle columns, it is necessary to run these columns on an instrument specifically designed to accommodate the pressures produced by smaller particles, as well as minimize band spreading. This cannot be achieved with conventional HPLC systems.

The ACQUITY UltraPerformance LC® System is a holistically designed solution that improves the performance and analytical data quality of chromatographic separations by considering all aspects of instrument and column design.

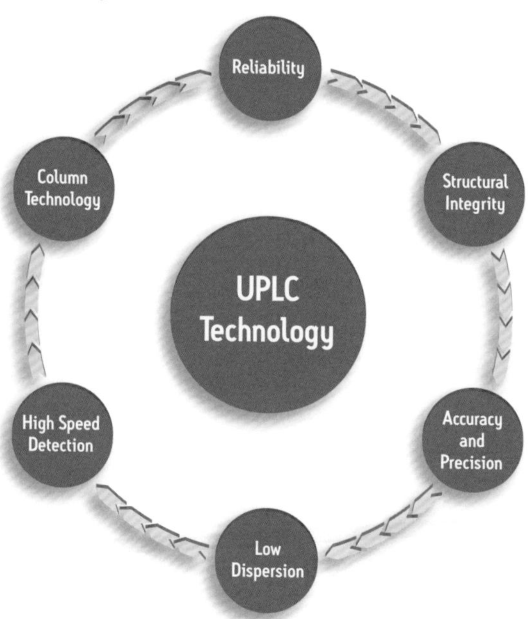

Figure 50: Holistically designed UPLC Technology.

It should now be clear that the key to UPLC separations is the combination of the instrument and column performance that allows chromatographers to fully realize and harness the power of sub-2 µm particle columns. This is achieved by minimizing band spreading within [intra-column] and outside [extra-column] of the column and being able to operate at the optimal linear velocities [and pressures] of these small particle columns [Figure 51].

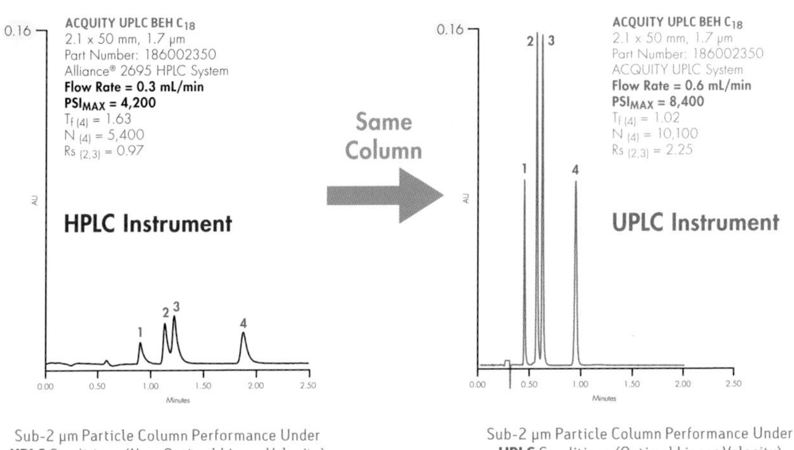

Sub-2 µm Particle Column Performance Under **HPLC** Conditions (Non-Optimal Linear Velocity)

Sub-2 µm Particle Column Performance Under **UPLC** Conditions (Optimal Linear Velocity)

Figure 51: The ability to operate in a fast, low band spread LC instrument capable of operating at the optimal linear velocity is crucial to realizing the performance gains of sub-2 µm particle columns. In this example four caffeine metabolites are analyzed using the same chromatographic conditions [except for flow rate as noted] on a fully-optimized, microbore HPLC instrument vs. a standard ACQUITY UPLC Instrument. The improvements in efficiency, resolution, peak shape and peak height illustrate the benefits of UPLC Technology and its holistic system design.

Maximizing Separation Power

In order to maximize separation power, one can combine the use of small particles with elevated temperature and elevated pressure to develop ultra-high efficiency separations using UPLC Technology.

Figure 52A is a single 150 mm long 1.7 µm UPLC column producing just under 40,000 plates at 90 °C. A second column was added in series to produce a length of 300 mm, resulting in a plate count of 83,000 [Figure 52B]. The full pressure range of the ACQUITY UPLC System is exploited by adding a third column in series, resulting in a 450 mm long UPLC column packed with 1.7 µm particles. As observed in Figure 52C, an efficiency of 121,000 plates is achieved in only 8 minutes.

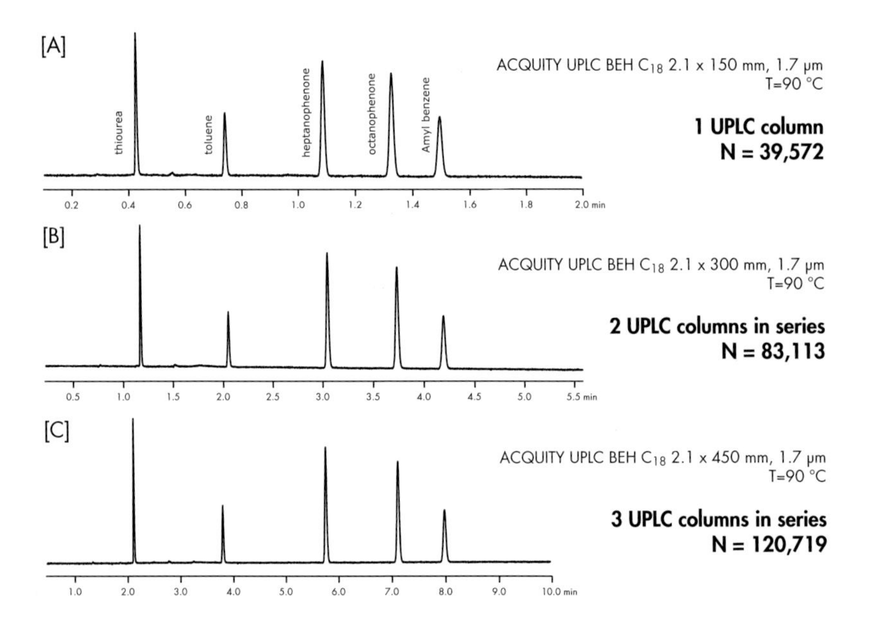

Figure 52: Combining elevated temperature with UPLC Technology to maximize plate count.

Gradient separation power can also be increased using the same logic. In this example, two 150 mm long, 1.7 µm UPLC columns [total length 300 mm] were linked in series to dramatically improve the information achieved for metabolite identification [Figure 53]. A peak capacity of over 1,000 was achieved within a one-hour run time. The ability to fully characterize this urine sample allows for the identification of metabolites of candidate pharmaceuticals, detection and identification of toxicity markers, and detection of toxins in therapeutic drug monitoring. In this case, resolving power, and therefore mass spectral quality, is dramatically improved, resulting in simplified data analysis, improved detection limits, and increased assay confidence.

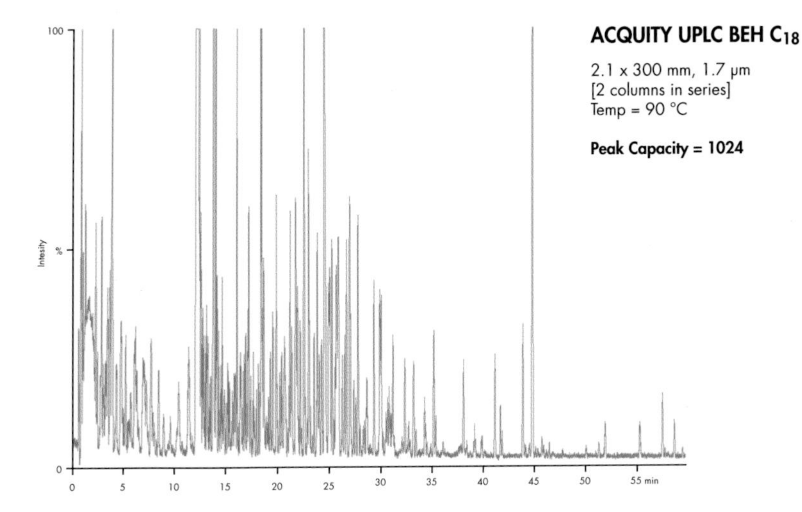

Figure 53: Combining elevated temperature with UPLC Technology to maximize peak capacity in a diabetic urine sample.

A Single System for HPLC and UPLC Separations

The ACQUITY UPLC System is designed to meet mounting organizational challenges to deliver products to market faster while maintaining or improving the quality of information. Since 2004, countless organizations have adopted UPLC Technology as a routine analytical platform, displacing conventional HPLC.

When adopting a new technology, it is important to consider its capabilities to meet existing and future demand. With UPLC Technology, one can invest in the future by utilizing a single system with the capability to optimally run sub-2 µm columns, and the ruggedness and robustness to run legacy HPLC methods. This means a single technology platform can be utilized independent of the separation needs, thus facilitating improved productivity by simplifying method transfer from site to site.

Figure 54 shows an example of the ability to run the ACQUITY UPLC System as a standard HPLC. This is the USP method for Excedrin [over-the-counter pain reliever] run on a conventional HPLC System [Figure 54A] and on an ACQUITY UPLC System [Figure 54B]. The chromatographic method, mobile phases, sample and column were simply moved from one instrument to the other. As a result of optimal system design, higher efficiency and sensitivity are observed for the same assay on the ACQUITY UPLC System without changes in selectivity or relative retention.

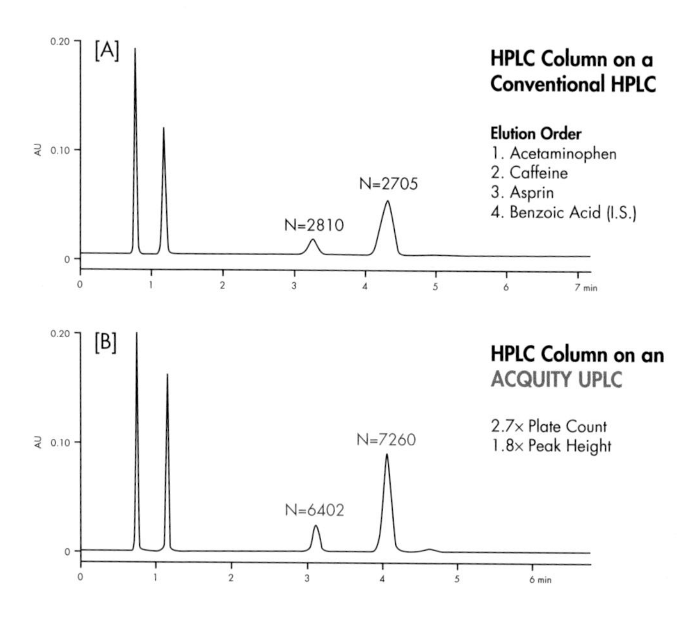

Figure 54: Performance of the ACQUITY UPLC Instrument as a standard HPLC. XBridge C_{18} 4.6 x 100 mm, 5 µm column run at 2.0 mL/min at 45 °C in a 73:23:3 water:methanol: acetic acid mobile phase. Detection at 275 nm, 5 Hz, digital filter = 0.1.

Conclusion

This technology primer is designed to provide the reader with a basic understanding of the chromatographic principles on which UPLC Technology is based. It is our hope that the reader now understands the significant improvements in resolution, sensitivity and speed that can be achieved for chromatographic separations by minimizing the band spreading contributions of both the instrument and the column. In addition to minimal band spreading, such a system needs to be capable of operating at the optimal linear velocity [and pressure] for small [sub-2 μm] particle columns. The ACQUITY UPLC System was designed to meet the present and future needs of separation scientists.

The theoretical efficiency and resolution gains described in this primer were predicted decades ago. For the first time, theory has been put into practice on a commercially available scale with the introduction of the ACQUITY UPLC System in 2004. Since then, many thousands of separations scientists around the world have embraced UPLC Technology and the practical benefits it can provide their organization. The ACQUITY UPLC System allows organizations to more effectively manage company assets by improving the quality of the chromatographic information and reducing the amount of time necessary to acquire this information. This drives organizational productivity by overcoming many of the challenges and bottlenecks present in the analytical separations laboratory. Besides providing obvious and significant consumable cost savings, shorter, more reliable separations also benefit the environment by requiring significantly lower quantities of organic solvents.

Interestingly, most leading instrumentation manufacturers, who originally downplayed the importance of a higher pressure-capable chromatographic system [citing safety, robustness, sample compatibility, etc. concerns], have now validated the need for such an LC platform by developing various higher pressure capable systems of their own. Although these systems do come with a fair bit of compromises [e.g., lower pressure limits, large band spread, limited detection choices, etc.] this unmistaken trend towards higher chromatographic performance does indicate that the separation science is still advancing.

References for Further Reading:

1. U.D. Neue, “HPLC Columns: Theory, Technology, and Practice,” Wiley-VCH [1997]

2. J.C. Aresenault and P.D. McDonald, “Beginners Guide to Liquid Chromatography,” Waters [2009]

3. P.D. McDonald, “The Quest for Ultra Performance in Liquid Chromatography: Origins of UPLC Technology,” Waters [2009]

4. M.P. Balogh, “The Mass Spectrometry Primer,” Waters [2009]